BLUFF YOUR WAY
IN
SCIENCE

BRIAN MALPASS

ℛℛ
RAVETTE BOOKS

Published by Ravette Books Limited
Egmont House
8 Clifford Street
London W1X 1RB
(071) 734 0221

Series Editor – Anne Tauté

Cover design – Jim Wire
Printing & binding – Cox & Wyman Ltd.
Production – Oval Projects Ltd.

The Bluffer's Guides ® is a
Registered Trademark

The Bluffer's Guides series is based
on an original idea by Peter Wolfe.

An Oval Project
for Ravette Books Ltd.,
an Egmont company.

For use of the brass microscope on
the cover, thanks are given to:
Arthur Middleton
(Antique Instruments)
12 New Row
Covent Garden
London WC2.

CONTENTS

Scientists:

WHAT SCIENCE IS

There is a very good reason for the inadequacy of the standard definitions of science – pay attention now – which is that, being concerned with life, the universe and very nearly everything, science is so fundamental it defies definition. (It also deifies definition, but we'll come to that.)

To illustrate the point we shall take just one example. Various worthies have had a shot at providing meaningful definitions of science, including Victorian thinker Herbert Spencer, who knew a thing or two, having coined such little gems as, 'The survival of the fittest' and, 'Billiards is a sign of an ill-spent youth'. But he was obviously having an off day when he produced, 'Science is organized knowledge'. On that basis, the telephone directory and any filing cabinet are science incarnate, which is such arrant nonsense it merits no further consideration.

So rather than pursuing a fruitless full-frontal assault, we propose to sneak up on science one veil at a time by telling you (a) what it's made up of, and (b) a few snappy facts about scientists.

If that sounds rather vague and and not very logical it will at least kill the most common misconception about science and scientists – that they are all about logic.

Think about the apple that fell on **Isaac Newton**'s head. If Newton had behaved logically he would have simply moved to sit somewhere not under a tree full of potential missiles. But he was a scientist, so he stayed where he was, scorning further disaster, and devised the Law of Gravity, which is so important it is usually spelled with Capital Letters.

The component sciences that make up the body

scientific are sometimes grouped into Life Sciences, Earth Sciences, and Physical Sciences. But let us be absolutely clear from the outset that what we are dealing with here is Natural Science. The Bluffer's Guide is an upright, straight-shooting and rather butch publication that has no truck with anything unnatural and will never be purveyed in a plain brown envelope.

As to the individual components, science appears nowadays in a bewildering variety of guises. Thus you can have molecular biology, genetic engineering, particle physics, and any number of even stranger hybrids, such as astrogeology, morbid anatomy, marine geophysics, pharmacokinetics, and behavioural neuroendocrinology.

When you encounter some exotically named subscience it can often be dealt with like a German adjective by simply analyzing it into its component parts.

But for an understanding of the scientific balance sheet you need only consider the big three:

Chemistry – The science of molecules, being concerned as it is with the study of material substances, and nothing whatever to do with what happens between man and woman.

Physics – The science of matter and energy and things smaller than molecules; in fact, increasingly concerned with things smaller than the smallest thing so far discovered.

These days, physicists tend to look smugger than most other scientists because it has been clear for many years that, in theoretical terms at least, even the big three sciences can be reduced ultimately to just physics.

One of the most famous of all chemists, **Linus Pauling** (*q.v.*) has admitted that fundamentally the whole of chemistry works on quantum principles. **P.A.M. Dirac**, the illustrious theoretical physicist, first blew the gaff when he pointed out that chemistry can be understood in terms of the interactions of the particles that make up atoms and molecules; and that those interactions can be expressed in quantum mechanical equations which could be solved if one had a big enough computer. However, chemists need not feel that they are an endangered species, for the average physicist is incapable of boiling an egg.

Biology – The science of living organisms. Biology used to be studied by those who felt that for some reason they had to make a token gesture towards the sciences but couldn't do maths. Anybody trying that lark nowadays is in for a severe shock because modern biology is riddled with statistical treatments which make Boolean algebra look like a Noddy book.

It is worth remembering, especially if you are a biology fan in need of cheering up, that **Aristotle** was primarily a biologist and he was the most influential scientist in history, at least for the length of time his teachings remained received wisdom, which is rather longer than anyone since.

All other 'ologies' should be distrusted, especially scientology. In passing, you can be dismissive of seismology on the grounds that in science as in other things seis doesn't matter. But cosmology is actually a pukka subject, being in fact nothing more than mathematical physics thinly disguised to avoid the off-putting effect that its real name would undoubtedly have upon the faint-hearted.

Astronomy is not really one of the foundation garment sciences. We mention it here purely for old-times' sake. For astronomy was the first science, and for millennia the most popular. Anyone who can tear himself away from the pleasures of the pit long enough on a starry night can do it. And in olden days those with brains as well as insomnia soon found that a good knowledge of astronomy was invaluable for scaring the living daylights out of the hoi polloi and keeping them firmly in their place through its use in calendrication, prediction and knowing the optimum times to put in bedding plants.

Beware the expression 'The Scientific School of ....' Many subjects boast 'scientific schools of' and the only thing that need be said about them here is that they have nothing whatsoever, in any way, shape, or form, to do with science. This applies particularly to Bridge and Boxing.

The rest are of limited interest, unless of course you feel the need to introduce some obscure science, real or imagined, so as to confuse the unwary.

Beware of heaping scorn on subjects such as palaeontology and archaeology, so-called 'soft' sciences because they purport to be non-mathematical. The scientific revolution has hardened them enormously.

There are any number of pseudosciences whose titles include the word 'science', not as an indication of genuine membership of the family of sciences but simply as an impressive suffix designed to lend an air of verisimilitude to an otherwise bald and unconvincing narrative. Examples abound, including 'Social Science', 'Christian Science', 'Management Science', and 'Political Science', from which it can be seen that they can not only be ignored, but openly despised.

How to Recognize a Scientist

Having defined science, the next thing is to learn how to recognize a real scientist. There is no 100 per cent reliable way, of course, but the following should provide a good working guide.

First of all, anyone carrying a copy of the *New Scientist* is not necessarily one. Most do read it on the sly but they keep quiet about it.

Anyone using the word 'boffin' is not a scientist, unless he speaks with a clipped intonation, as in "huntin', shootin', and fission".

Anyone wearing a white coat probably is a scientist, unless accompanied by another similarly attired and carrying a straitjacket. The presence of a variety of writing implements protruding from the top pocket is a pretty reliable indicator that one is dealing with the real McCoy too. Naturally, other people do wear white coats, such as hospital doctors and those who flog toothpaste on television, but theirs are not usually covered in acid burns and sinister stains.

Be on the lookout for galloping tabulaphilia, which may be defined as an unhealthy attachment to blackboards, and is a surefire pointer to physicists. For while one physicist in isolation can talk to a lay person, admittedly with difficulty, two physicists can only converse with each other in close proximity to a blackboard, on which they feverishly scrawl cabalistic symbols. If there isn't a blackboard handy, they will happily use your living-room wall.

Anomalous behaviour in lavatories can be a give-away too. If you see someone wash his hands *before* having a pee, he is indelibly marked as a chemist.

But the real acid test is to look at the eyes, which can be read like a book with a little practice. The eyes of the true scientist show nothing but quiet desperation. The reasons for this are twofold:

1. The fact that it is never possible to prove scientific theories correct. They can only be demonstrated to be incorrect. This somehow weighs heavily on the collective scientific subconscious.

2. The fact that the scientist knows with total certainty that his work is the most important of all, being concerned with the very fabric of the universe itself rather than the ultimately inconsequential activities of humankind, and yet his profession ranks in the social scale somewhere between street-sweepers and politicians, at least in Britain.

If all else fails, a study of the subject's grammar, syntax and speech patterns will often give the game away. Scientists say things like, 'It all depends what you mean by...', 'To a first approximation...', 'Within the limits of experimental error...', 'Define your terms...', and 'Orders of magnitude' (which is nothing to do with sergeant-majors shouting commands). All are simply devices to give the speaker time to think what to say next.

Scientists tend to talk in an odd stilted way, using verbal versions of written abbreviations. For example, take 'for example'. The scientist pronounces that as 'ee gee' like an expletive uttered by a naturalised American of Lancastrian origin. And instead of saying, 'that is', the scientist says 'aye eee', like the death cry of the baddie in the comics to which all scientists are secretly addicted.

Finally, and most remarkably, scientists speak in

the numbered and lettered paragraphs and subparagraphs much used in learned journals. The real expert can actually fashion verbal footnotes complete with aural asterisks but the novice is advised not to attempt this without a great deal of practice.

In assigning relative weighting to the different speech indicators, it is wise to regard a single 'to a first approximation' as equivalent to two of any of the others. Thereafter, simply perm any four from five in reaching your final assessment.

Once you have detected the presence of the genuine article you can identify the scientist's own specialised subject. Provided you do that and steer clear of it, the whole of the rest of science is fair game, since science has become so esoteric and specialised that, out of his own narrow bailiwick, the scientist speaks as a child, and may be shown no mercy.

The way to identify the subject's specialised field is by the jargon, which is denser and more impenetrable than any thorn thicket. Most is so arcane that the layman cannot even hazard a guess at what the words are, let alone what they mean, but there are a few exceptions, e.g: 'quarks' (especially if pronounced to rhyme with 'porks'). Your interlocutor is a particle physicist who probably works at the European Centre for Nuclear Research which is buried underground somewhere on the Continent. (The fact that this is called CERN has nothing to do with a sensible desire to have an easily pronounceable acronym, and everything to do with the mulish insistence on the part of the French to render everything in sight into their native tongue.)

Oddities like 'proton', 'photon', 'neutron', 'lepton', 'hadron', 'muon', 'gluon', 'meson', 'pion', 'axion' and 'klingon' and peculiar modifiers like 'upness', 'down-

11

ness', 'strangeness', and 'charm' refer to some of the ever smaller and ever weirder particles that are being uncovered daily and immediately label the speaker as a refugee from the atom smasher.

'Poly', 'oligo' and 'mono' (unless they have 'glot', 'psony' and 'tonous' tacked on to them) and methyl α-cyano acrylate, no matter what it has tacked on to it, (which is probably quite a lot since it is the active constituent of Superglue) all unambiguously label the speaker as a polymer scientist.

These unhappy beings were until recently the stars in the scientific firmament, before 'greenness' (the other sort, not naïvety) took a hold and 'plastics' became a dirty word. Beware, by the way, the term 'free radical', which may identify the speaker as a polymer man, but may also indicate that you are dealing with an escaped revolutionary.

If you hear the word 'enema', do not turn away too hastily in the belief that you are about to be treated to a sick-making blow-by-blow description of the speaker's last operation. You may have misheard 'NMR', short for 'nuclear magnetic resonance' which is arguably the most powerful instrumental technique available to the chemist and a non-invasive aid to diagnostic medicine.

The odd-sounding expression 'p-holes' may well indicate an old-fashioned solid-state physicist. (The position of the hyphen is vital here. A solid state-physicist is a politically correct government scientist.) Try muttering the name Dirac as a confirmatory test. If the subject rises to his feet and assumes a look of dog-like devotion, you've nailed him. Unfortunately the homonym 'pee-holes' can cause some confusion but does not mean the speaker is not a scientist.

The list is endless for the growth in science has

been so rapid that 90 per cent of all the scientists who ever lived are alive and working today – except in the UK where like everyone else they are simply alive.

Provided you take precautions as described above, the professional scientist is a legitimate and not particularly difficult target. The person who must be avoided at all costs is the polymath (or renaissance man) who has taken the trouble to read up on the subject. Such people are extremely dangerous because they are articulate and cannot be thrown off the track by retreating into jargon, real or imaginary.

Recognizing these walking mine-fields is tricky because they come in all shapes and sizes. We are told by those who deal in psychic phenomena that these people have an aura which is black, so that they look as if they are standing under their very own personal thunder cloud. Later we shall be giving you more hints on how to stop them raining on your parade.

Succeeding in Science

The trick to succeeding in science is simple really. You have to become the world's greatest living expert in some tiny facet of your chosen subject. The successful scientist's motto could well be, 'If you've got a niche, scratch it.'

The bit you choose does not have to be interesting or important or have any great practical use. In fact the more tedious and trivial it is the better, because then the chances of someone horning in on your patch are correspondingly less.

If some busybody does find a use for what you do, so that all sorts of undesirables row in, you can

13

always up sticks and find another hidyhole in the reef of science to back yourself into. We personally had to go through this tiresome manoeuvre twice before finally giving up in disgust and leaving science altogether.

It is not always as easy as you might think to tell when a scientist is successful. But one sound practical indication that your efforts are being rewarded by a modicum of success is that other scientists start to hate you. If your particular branch of science should suddenly become topical, you might get on television, but you need to develop a speech impediment to be really successful and not everyone is willing to go that far.

The surest sign of incipient success is that you cop an award. At all levels of science there are more awards than you can shake a stick at, doubtless to compensate for the lousy pay.

The most prestigious are the annual Nobel prizes for Physics and Chemistry. These represent the penultimate symbol of success in science but since the winners often receive the award many years after the publication of the work concerned it is not always a useful guide.

Some Nobel prizes are sympathy awards. The last example was probably the American physicist **Norman F. Ramsey**, whose award in 1989 was in recognition of his life's work rather than any specific breakthrough. Ramsey's achievement in winning the prize was especially meritorious when one considers that he has completely the wrong sort of name. Something polysyllabic, with a touch of the exotic and mysterious about it, is normally required for success. Ashkenazic names are particularly favoured.

Many scientists regard the Nobel prizes as being

contrary to the whole spirit of science; the selfless, dedicated search for truth and beauty in the world without thought of material reward. Others point out that it is intrinsically suspect for any group of people to decide subjectively and arbitrarily whether one piece of scientific work is 'better' than another, even those as well-meaning and meticulous as the Scandiwegian academies which dole them out.

Despite this, we shall be making liberal use of success in the Nobel prize stakes as a rough measure of scientific excellence. For the essence of science is quantification. As **Rutherford** said, you may think you know a subject but until you can measure it and calculate it, your knowledge remains of a vague and unsatisfactory kind. Of course, Rutherford did hail from New Zealand, but nevertheless he did have a point.

In science there is no escape into the comfortable vagueness of words, nor indeed into the uncomfortable vagueness of the numbers used by accountants with no estimates of the degree of uncertainty to be attached to them.

Perhaps the ultimate sign that you have arrived in science is to have a unit of measurement named after you, although it can take two hundred years for recognition to arrive, which lessens its value to the individual concerned. Examples include the coulomb, gauss, ohm, oersted, volt, newton and twaddell.

Even the most cursory inspection of the partial list of units above will illustrate one of the most subtle laws of science which is that eponymic units of measurement only sound genuine when foreign names are used.

The only thing that outranks having a unit of measurement named after you is to have your moniker tacked to a law, theory, rule or principle. Or better still, a scientific phenomenon. The eponymic law applies with equal force here. Take for example, Raman spectroscopy, redolent of Eastern promise, or Cerenkov radiation, ripe with Slav mystery. Contrast them with the Appleton layer, which directs the mind irresistibly to a type of domestic chicken, or the closely-related Heaviside layer with its unpleasant connotations of cellulite.

Sad to say, but some scientists go overboard in their pursuit of success and this applies particularly to the acquisition of terminal letters denoting their degrees. Initially, the reasons for this title hunt are entirely practical. In science, you cannot so much as wash a test-tube without possessing the appropriate label.

So anyone wishing to make a career in science must collect tags. The trouble with starting out on this path is knowing where to stop. There are ordinary degrees, pass degrees, honours degrees, and higher degrees. Then after a brief detour through associate-ship, membership, and ultimately fellowship of your professional body, and possibly election to the Royal Society, it is back to the degrees with the award of Doctor of Science, which is based on the weight in kilograms of the papers you have published.

Finally, you start to collect honorary degrees, the more the better. It is quite possible to kill yourself by degrees. Mind, it is not invariably fatal. We seem to recall that **Lord Kelvin**, the famous physicist, used to be the acknowledged champion of terminal letters. It was said of him that he was entitled to more letters after his name than any man in the Commonwealth, on which the sun never set, and he lived to be eighty-three.

THE GROWTH OF SCIENCE

The growth of science this century may have been explosive but for all the rest of recorded history progress was very slow owing to a variety of factors, chief among them being:

1. Mathematics

Normally the only philosopher to whom scientists pay any real heed is Karl Popper. However, in this instance points may be earned by quoting Wittgenstein. It was he who said that the limits of our language set the boundaries to our world. Since mathematics is the language of science, any inadequacies in it undoubtedly held back many of the peoples of antiquity, from the Babylonians who counted in 60s to the Greeks who had no convenient symbol for zero, making multiplication by ten rather tricky.

The situation was only really resolved when the Arabs on their trading perambulations picked up the necessary wrinkles from the Hindus.

2. Attitudes

The Greeks treated science largely as an intellectual exercise and took a dim view of anyone finding practical applications for it, other than the military. They felt that it ought to be possible to elucidate the laws of the universe by the sheer power of the mind.

The Romans seemed to feel that they got along pretty well without science so why bother. Again, it was the advent of the Arabs, bringing a welcome eclecticism and pragmatism that marked the climacteric between the cerebral science of antiquity and today's observational variety.

The Scientific Method

This is the big one, and may have been preying on your mind. You are not alone. People often pant up to us as we are browsing through the fruit and veg in the supermarket and demand, 'For god's sake, Doc, I have to know. What's the difference between a scientific law, and a mere theory? And how does the humble hypothesis fit in?'

In the interests of pursuing our quest for the perfect avocado, we are sometimes tempted to dodge the column by replying laughingly that anything by Newton is a law, while anything by lesser beings such as Einstein, A., is a theory. But discerning the anxiety on the faces of our earnest questioners, we ruthlessly suppress the desire to dissemble, and come clean.

We tell them that the differences become clearer if you consider for a moment the scientific method. From observations, or sometimes simply from the effect of eating curry too close to bedtime, you dream up a theory.

You then use that theory to hypothesise, to make predictions, and you test your hypotheses by means of experiments designed to prove them. (Or, if you are a follower of Popper, to disprove them.) If the results of the experiments agree with the predictions, then the theory is pronounced good – so far.

Sooner or later someone comes along with some

more extensive or more accurate observations which the theory is unable to account for. The theory is then tinkered with to make it fit the facts.

This iterative process may be gone through several times with the theory growing into a more and more unwieldy edifice until someone sits down with a clean sheet of paper and comes up with a brand new theory that accounts for the observations in a simpler and more elegant way.

Then Occam's Razor is brought into play. This says that faced with two equally tenable explanations, one complex, the other simple, you go for the latter as a Jack Russell goes for a rat.

To show that science is no respecter of persons, precisely this sort of thing happened to Newton's Law of Gravity, which eventually had to bow the knee to Einstein's version, the general theory of relativity.

Perhaps it is simply that having seen so many so-called laws bite the dust of experimental fact, we are reluctant nowadays to confer the status of law so lightly. Take the case of QED; not *quod erat demonstrandum* but quantum electrodynamics, of which the late, great Richard Feynman was high priest. This has been undoubtedly the most successful theory of all time, its experimental verifications agreeing with theoretical predictions not approximately but to the nth decimal. Yet it has never been ennobled but remains a mere theory.

In modern science, experiment and observation are king. A law or theory, no matter how ingeniously fashioned, no matter how elegantly phrased, is like the present-day football manager – only as good as the last result.

Who is Top at Science

The answer to this seems to depend crucially upon which period in history is being considered. At one time or another various nations have perched atop the scientific dungheap only to be knocked off by someone else.

1. The Arabs

It should already be clear how a collection of nomadic traders became the first kingpins of science as we know it. Few individual Arab scientists survive as household names, but they left a rich legacy of scientific terms, especially those starting with 'al...' Cite, inter alia, (which is Latin for 'among other als'), alchemy, alembic (from al-anbiq, a distillation apparatus), alidade (a revolving rule used in surveying), alkali (from al-qili, the ashes of the saltwort plant, now used of highly corrosive substances good for removing things, especially skin), and finally, stretching a point, algebra and algorithm, both something to do with sums.

The golden age of Arab science was more or less contemporaneous with our Dark Ages. Ever since they discovered they were sitting on most of the world's oil they have paid others to do all the distasteful things, like brain surgery and science, for them.

2. The French

After the Arabs (not immediately after you understand; this is only an overview), it is necessary to mention the French. The rollcall of gallant Gallic scientists makes impressive reading. Who hasn't heard of **Pasteur** (*q.v.*), **Gay-Lussac**, (unfortunately-named but blameless), **Lavoisier** (*q.v.*) and **Le Chatelier** (whose Principle is the most elegant statement yet devised of the cussedness of inanimate objects).

3. The Germans

Just consider the following list of Teutonic scientific Nobel laureates: **Hans Albrecht Bethe** (1967), **Konrad Bloch** (1964), **Max Born** (1954), **Hans Georg Dehmelt** (1989), **Gerhard Herzberg** (1971), **Polykarp Kusch** (1955), **Albert Michelson** (1907), and **Otto Stern** (1943). What a team.

Lots of countries would have been proud of them, and indeed lots of countries were because, although they were all born German, they were naturalized citizens of other nations when they won their awards, as was **Albert Einstein** who was worth, at least in terms of autographs, all the rest put together.

Even today the Germans come second only to the USA in the number of scientific Nobel prizes pocketed.

4. The Americans

Despite a distressing inability to spell any scientific term containing the letters 'ph' in combination, the Americans are top of the Nobel league table by a mile, and we do mean a mile, since they stubbornly refuse to go metric.

The USA seems to have fielded most of the Jewish scientists who fled Nazi Germany between the wars. At a rough count, getting on for 20 per cent of US

Nobel prizes in science have been awarded to naturalised citizens.

Above all, the go-ahead Americans are willing to spend money, which makes them a sink into which drain the brains from the rest of the world.

5. The British

Flying in the face of the low esteem and monetary rewards that seems to be the scientist's ineluctable lot in the UK, the British amazingly continue to turn out good scientists and to do good science. In terms of Nobel prizes per head of population the United Kingdom, in total only narrowly behind Germany if our calculations are correct, must be about top of the charts.

Like the USA, it too has attracted a fair number of refugee scientists, desperate people fleeing not only the Nazi terror but the more subtle privations of the Antipodes. Off the top of the head we can think of at least two UK Nobel laureates of New Zealand origin.

6. The Women

The recommended approach to the issue of women in science, as in so many other things, is to counterpunch, or better still hit on the break if you can do it on the referee's blind side.

If grappling with a feminist, say brightly, "Well of course women are good at science. Look at the women who have won Nobel prizes. Let's see, there's Madame Curie. And then there's ...." At this point, you allow your voice to trail away uncertainly into "Oh, crumbs!" If you are dealing with a male chauvinist you must hope that your antagonist is sufficiently ignorant even to be aware that, for instance, there were two Mesdames Curies. **Marie** actually won two Nobels, one for chemistry, the other for physics, and her daughter **Irene** won one. As if that wasn't enough, they both did it despite the almost insuperable disadvantage of being married to Frenchmen.

Among the other female recipients of the award of whom no-one has heard is a Mrs. **Maria Goeppert Mayer**, who nobbled one for nuclear shell theory and who was heard to explain spin-orbit coupling in terms of the Viennese waltz. Rather better known, although still not exactly a name on everyone's lips, is **Dorothy Hodgkin** who worked out the structure of vitamin B12.

Then there was the spiky but brilliant British physical chemist **Rosie Franklin**, the unsung heroine of the elucidation of the structure of the double helix of DNA, who should have won one but didn't, lacking some of her male colleagues' gift for self-publicity. In the same galère was **Lise Meitner**, a German Jew who flew the coop in 1938. Although a brilliant physicist, she clearly didn't have much of a sense of direction because she fled to Sweden and didn't leave there until she retired. It was she who coined the expression 'fission'.

There are still others – **Rita Levi-Montalcini** (neurologist, 1986), **Barbara McClintock** (geneticist, 1982), **Gerty Radnitz-Cori** (biochemist, 1947) and **Rosalyn Yalow** (physicist/endocrinologist, 1977) – if you cheat slightly and count those scientists who were awarded the Nobel prize for Medicine.

Whither Science?

At various times in the long and chequered history of science, eminent practitioners have surveyed the body scientific, made symbolic dusting motions with their hands, and pronounced with gloomy satisfaction that that was just about it. And this was invariably the signal for a series of major discoveries that turned the whole thing on its head and set it off in new directions, hitherto undreamt of.

The last such occasion was towards the end of the 19th century, when many scientists firmly believed that no new advances in physics remained to be made.

Quote **Albert Michelson**, the first American science Nobel laureate, whose experiment to measure the velocity of light in the direction of the Earth's path in space and at right angles to it (in an attempt to detect the mysterious 'ether') is perhaps the most momentous single negative experiment ever carried out.

Michelson remarked that it seemed probable the grand underlying principles of physical science had been firmly established and that further advances were to be sought chiefly in the vigorous application of those principles to all phenomena.

He was not a lone fuddy-duddy. The view he articulated was widely held. Naturally, very soon after he uttered it, the completely new phenomenon of radioactivity burst on to the scene.

And of course even as he spoke the teenage Albert Einstein was beavering away, fashioning the bombshells that he was to lob into the complacent pond of science with the publication in 1905 of no less than *four* devastating discoveries in physics. Ironically, one of them, the special theory of relativity, explained satisfactorily for the first time the results of Michelson's own famous experiment.

So if ever a truly eminent man of science over the age of fifty-five pronounces with great conviction that something is impossible, it is a safe bet that he will be proved wrong shortly. And should another equally great man of science rush to endorse his colleague's assertion, you are advised to mortgage the family dwelling and hasten to a betting shop to invest the proceeds at two to one and await the imminent announcement to the contrary.

One of the finest examples involves no less a person than Rutherford, the father of the atom, who injudiciously remarked that any idea that the atom could ever provide useful power was total rubbish. Inevitably, the first nuclear reactor was in operation comfortably within ten years of his pronouncement.

Nowadays the role of Jeremiah and harbinger of doom has been taken over by **Stephen Hawking** (*q.v.*), who is fond of saying that the end of theoretical physics is in sight. But Hawking is renowned for his puckish sense of humour and is surely jerking our collective string, or super-string (*q.v.*) to be bang up-to-date, when he says it.

So despite the enormous growth in science already experienced in this century you can confidently advise all and sundry to expect no slowing down in the torrent of exciting developments.

Here are a few to pontificate on over the wafer-thin mints.

SCIENTIFIC ISSUES

In examining any issues confronting science you are immediately faced with an embarrassment of riches or a plethora of problems, depending on which way you look at it.

The Big Bang (Physics)

The idea that the universe was conceived in one gigantic primeval bang still represents best current thought. There used to be a rival theory called Steady State in which new matter was postulated to be popping into existence all over the place the whole time, but that was routed some time ago.

According to Big Bang, the whole of the known universe once existed as a singularity, a vanishingly small, mind-blowingly dense concentration of matter that made the average tin of sardines look thinly populated by comparison. This singularity then took it into its head to explode. Surprisingly, not even the IRA has admitted liability so no one knows why it blew up, although Stephen Hawking has come up with a rather neat way of avoiding the question.

Since Big Bang, the Universe has been expanding and the knotty question is whether it is 'open' or 'closed'. That is to say, will it continue to centrifuge ad infinitum or will it eventually stop flying outwards, pause, and then start to collapse in on itself, eventually disappearing up its own singularity once more.

To decide between these two, on the face of it, equally unattractive outcomes you must, as always, measure and quantify. Talk in terms of round billions.

This is quite a common technique to enable one to mention the unimaginable without gibbering. Economists and politicians deal with budget deficits in a very similar way.

What is going to be crucial in settling whether it is plan A or B that is being put into operation is the amount of matter in the universe. If there is enough of the stuff it will be able to arrest the expansion; if not it won't. First estimates have not been encouraging. Not only is there not enough, the size of the shortfall is on a scale more normally associated with the National Debt.

In fact, there seems to be at best only ten per cent of the quantity of matter needed to stop the runaway. However, all may not be lost, for there is now the concept of 'dark matter'. This is the stuff which, unlike normal matter that is visible in one or other part of the electromagnetic spectrum to our instruments, for reasons of its own just sits out there not emitting anything we can pick up readily, although its presence can be inferred from such things as gravitational image splitting.

This intriguing concept has thrown the whole issue back into the melting pot. It means that, along with everyone else, you will have to watch this space.

The Big Bang is believed to have taken place some 15 billion years ago and even if everything were to come to a grinding halt in the next ten minutes or so it might reasonably be expected that the process of collapse, by symmetry, ought to take a further 15 billion years. So there is definitely no cause for alarm.

What is causing some agitation is Hubble's constant, named after the American astronomer **Edwin Hubble**. Since the constant is related to the age of the universe, its accurate measurement is crucial.

Those who attempt to compute it fall into two groups, the High-hubblers and the Low-hubblers. The two camps argue more viciously than Swift's Big-Endians and Little-Endians ever did. At the moment the High-hubblers seem to be gaining the upper hand and if they prevail it would mean that the oldest galaxies are older than the universe itself, casting doubt on the whole concept of Big Bang. Hubble bubble, toil and trouble.

GUTS (Physics)

The other great issue in this neck of the woods is the all-singing, all-dancing 36-dimensional theory of absolutely everything. Some theoreticians, notably the redoubtable Hawking, seem to believe that it will eventually be possible to combine into one Theory of Everything (TOE) or Grand Unified Theory (GUT) the four theories presently required to explain how the universe works.

As any boy scout will tell you, you never know when a piece of string might come in handy and this applies just as well to new physics as to old trousers. We are talking relativistic, quantized super- (in the sense of super-symmetric) string naturally. String theory holds that all the matter and energy particles hitherto thought of as points are better visualised as tiny vibrating strings.

In the opinion of many who ought to know, this unlikely-sounding premise seems to offer the most promising jumping-off point for the formulation of an eventual GUT.

We do not propose to go into this in any detail as it shares along with vintage port the alarming tendency to start a dull throbbing behind the eyeballs. We believe history will show that the search will demonstrate an interesting relativistic effect. No matter how fast we approach the GUT, it will always be 20 years off.

This is entirely appropriate, for Einstein himself spent much of his declining years biting on this particular piece of granite.

New Materials (Chemistry)

Chemistry must produce – to order – new molecules with specific properties. And it does. Take the recently discovered buckminsterfullerenes or buckyballs, so named because they constitute a family of compounds of the element carbon whose structure bears a fanciful resemblance to that of the eponymous geodesic domes.

Their existence demonstrates that carbon still has a few tricks up its sleeve because until they came along it had only ever been found in two allotropic forms: diamond and the immensely slippery graphite. Already there are indications that buckyballs may have all manner of uses, ranging from high temperature superconductors to aids to new synthetic routes.

The Origins of Life (Biology/biochemistry)

Chemists have been very fond over the years of producing primeval soup. This is nothing whatever to do with refectory food in universities, but refers to experiments which involve taking a gaseous mixture simulating the early Earth's atmosphere and exposing it to sparks, to simulate lightning, and UV radiation.

Eventually, the water placed in the apparatus to dissolve the products of the reactions is found to contain complex mixtures of molecules which are the precursors of proteins, carbohydrates, and nucleic acids. Under certain conditions even protein-like molecules with some of the catalytic properties of enzymes have been produced, as has adenosine triphosphate (ATP), the energy store of cells. So the raw materials of life as we know it can be shown to be quite capable of being generated.

However there is still no fully satisfactory explanation for how the gigantic leap was made from this promising but inanimate gunge to a living entity capable of reproducing itself and evolving further. Despite some marvellous ingenuity involving everything from chemical reactions in comets to adsorption on clays, the vital step remains obscure. Game on.

Energy (Free for all)

There is a pressing need for an energy-generating process that is both inexhaustible and truly green. The physicists have been working on their entry in this field, nuclear fusion, for donkey's years. In principle, the fusion of deuterium, or heavy hydrogen, atoms to form helium with the liberation of vast amounts of energy according to Einstein's equation, should be the answer.

But the technical problems of holding the precursor nuclei in sufficiently close contact for long enough and at a high enough temperature (we are talking around a hundred million degrees here) have so far proved intractable.

There was some excitement when a couple of electrochemists, **Pons** and **Fleischmann**, said they had found a way to produce energy almost literally in a test-tube, the so-called 'cold fusion', by adsorbing the deuterium atoms on to a palladium cathode, which was thought to bring them into intimate contact virtually indefinitely. It turned out that their results were not reproducible, and the establishment has turned its back on the inventors, but the idea is so intriguing this one will run and run.

The Brazilians thought for a while they had cracked the conundrum by using two of their most abundant assets, sun and soil, to develop the fermentation of sugar cane to produce ethyl alcohol which could then be used to power the internal combustion engine.

The only snags seemed to be that the fuel was rather more corrosive than the petroleum-based products it was designed to replace, and that all the cars in Rio had bloodshot headlights.

However when the overall energy balance sums were done more carefully it turned out that the process consumed more energy than it produced.

A promising current candidate is the gasification and subsequent burning of coppiced willow wood, a material older than recorded civilization, which may show that even in science there is nothing new under the sun.

WHO'S WHO IN SCIENCE

The great, the good and the gruesome in science can be divided into two categories.

1. The everyday, run-of-the-mill, ten-a-penny geniuses about each of whom we will impart everything the bluffer needs to know in one paragraph or less.

2. The supermodels or Cindy Crawfords of the spectroscope, who have not so much edged back the frontiers of knowledge as seized them by the scruff of the neck and dragged them kicking and screaming over the far horizon. To these we shall devote rather more space.

You will notice that our choice is heavily concentrated in the summilexic, those whose surnames start with A-M. And we only venture into the fundilexic second half of the alphabet as far as P.

The cynic might assume that the reason for this is that we are working from a part-work encyclopaedia and are still awaiting the later volumes. Not so; it is simply that our maternal grandmother, who smoked like a chimney before it became unfashionable to do so, was a generous soul who used to donate to us the '*Great Men of Science*' cigarette cards that came with the packs. As we were feverishly awaiting the letter Q in the series, she switched brands.

Archimedes

Though far and away the greatest scientist of antiquity, the dates of his birth and death are not known precisely, but since both took place in Sicily it is hardly surprising.

He is reported to have leapt out of the bath shouting 'Eureka!', meaning 'I've got it', now known to have signalled his discovery of Archimedes' principle which, as everyone knows, states that when a body is completely immersed in water the phone rings downstairs.

He gave a brilliantly precise explanation of the lever and said, 'Give me a fulcrum and I'll move the world.' (For years we thought 'fulcrum' was Latin for laxative.) He also devised the Archimedes screw.

Aristotle (384-322 BC)

Aristotle doubled as a scientist and philosopher, one of the earliest known examples of moonlighting. He had a finger in so many pies it is quicker to say what he wasn't involved in. He is said to have been one of the two great intellects of ancient Greece. His more fervent admirers will tell you he was both of them.

His fame rests securely on his view that theory should be based on observation, and he was a compulsive collector of facts. Regrettably, the 'facts' he accumulated so diligently included unsubstantiated rumour, speculation, hearsay and folklore. Such was his hold on science that the reluctance of later generations of scientists to challenge his teachings actually impeded progress. For example, Aristotle believed that the blood didn't circulate in the body but simply ebbed and flowed. It took 2000 years before William Harvey plucked up the courage to prove otherwise.

Nils Bohr (1885-1962)

By the time this physicist was born scientists could at last afford christian names. Bohr is best remembered for his contributions to quantum phenomena, which even he admitted to finding shocking, and the principle of complementarity, later particularised by **Heisenberg**, which basically holds that you can't have your cake and eat it.

Nils Bohr-ed for Denmark and was from the Marlon Brando school of oratory (mumble, mumble), which is a profound pity. God knows what he might have told the world if the world had only been able to make out what the devil he was saying.

Robert Boyle (1627-1691)

Celebrated Anglo-Irish chemist. He was a great debunker of Aristotle, always a good sign. He did his best work on gases, hence Boyle's law. (No, not 'A watched kettle never Boyles', the other one about the volume of a given quantity of gas being inversely proportional to its pressure at constant temperature.) His work is said to have inspired Dean Swift to write *Gulliver's Travels*.

Nicolaus Copernicus (1473-1543)

One Polish priest who didn't make it to the Vatican. As an astronomer, he was among the first to practise accurate observation and description, and craftily did not publish his great work on the movement of the celestial bodies (in which he proved heretically that the sun, not the Earth, was at the centre of the solar system) until the year of his death. If the need should arise to use his name in conversation, care should be taken to place the stress on the second syllable, rather than pronouncing it 'Copper knickers', as generations of schoolboys have.

John Dalton (1766-1844)

This British scientist lived for many years in Manchester and not surprisingly took a keen interest in meteorology, especially with regard to monsoons. He studied the composition of chemical compounds and deduced, from the constancy of the proportions of the different elements they contained, that they consisted of discrete atoms joined together to form molecules. Due to the Manchester climate, he had a delicate constitution, hence the expression, 'Dalton's weakly'.

Charles Darwin (1809-1882) and Alfred Russel Wallace (1823-1913)

Darwin originally went to university in Edinburgh to study medicine, the family trade, but packed it in when he found he couldn't stand the sight of blood. He only landed the job, acting unpaid, of naturalist and port-passer on the *HMS Beagle*, because strings were pulled.

During the 5-year voyage he learned by doing, writing reams of notes, largely to take his mind off his chronic seasickness, and sending back tons of samples on the numerous stopovers. Subsequent analysis of these hardened his growing doubts about the fixity of species and led him to his theory of natural selection.

Darwin was chary of actually publishing his unorthodox views, suspecting correctly that they would not go down well with the Church. So he happily played with his ideas in private for over twenty years. Then Wallace, an obscure British naturalist who got his best ideas while suffering from malaria, sent him a paper which demonstrated that he had arrived independently at the same conclusions. Friends and colleagues rallied round and persuaded Darwin to read a joint paper with Wallace.

Your position as to whether Darwin or Wallace should be regarded as the originator of natural selection should as usual be determined by tactical factors. Whatever position your interlocutor takes, adopt the other stance.

Darwin never had a unit of measurement named after him but he did give his name to the town in Australia. (They changed it to Palmerston later but then decided that was even worse and changed it back again.) Darwin will always be remembered as the first man to advance a viable theory to explain why human beings are ticklish. Wallace had the last laugh by living to be ninety.

Albert Einstein (1879-1955)

Einstein was born in Germany, spent some of his formative years in Italy, became a Swiss citizen in 1900, moved back to Berlin in 1914, resumed his German nationality, and pushed off to Princeton in 1933, being deprived of his German citizenship in so doing but not becoming a US citizen until 1940. Does that indicate confusion or what?

He seems to have been in his time a cryptoagnostic with pantheistic leanings who frequently interpreted the actions of God (e.g. 'God is subtle but not malicious'), a liberal who suggested to Freud in a famous correspondence that humankind has a basic desire for antagonism and destruction, and a pacifist who urged the president of the United States to develop nuclear weapons.

He spent a large part of his working life in a forlorn pursuit of a 'unified field theory' which was doomed to failure from the start because he stubbornly rejected quantum mechanics and didn't really know enough about what it was he was trying to unify. He became a committed Zionist who turned down the presidency of Israel.

After achieving the fame he never sought or really understood, Einstein went through life looking permanently bewildered, like an eccentric uncle not quite mad enough to be kept locked in the attic, and it is this that you should home in on rather than the man's contribution to the sum total of human knowledge, immense as that undoubtedly was.

His early career got off to a faltering start but all that was forgotten with the publication of his first scientific paper, *Über die von der molekularkinetischen Theorie der Wärme gefordete Bewegung von in ruhenden Flüssigkeiten suspendierten Teilchen* which not only explained Brownian motion but also set new standards for the length and incomprehensibility of titles.

Eventually, he came up with the piece of work with which he will be for ever associated, relativity, a concept so devastatingly new that it had just about everyone foxed. **Sir Arthur Eddington**, the British astrophysicist who was a brilliant scientist in his own right, made something of a life's work out of explaining relativity in plain English. He was asked once by a journalist if it was true that only three people in the entire world understood relativity. Eddington thought about it then asked who the third one was.

To this day, despite Eddington's best efforts, few people truly comprehend what relativity is all about, and indeed one of the pitfalls confronting the scientific bluffer is that sooner or later some clever dick will fix you with a beady gaze and demand that you explain it.

Resisting the temptation to move off, try murmuring something to the effect that the special theory, of course, changed for ever the preoccupation of physics from mere events to the relationship between events and the observer. Don't forget that little phrase 'of course', which subtly conveys the clear impression that the questioner is rather dim for having to ask in the first place.

But the most telling put-down you can inflict upon the questioner is to explain condescendingly why Einstein was interested in relativity in the first place, something that we can now reveal here for the first time. His study of the subject was a marvellous example of prescience because it enabled him, after getting divorced from his first wife and marrying the widowed daughter of his late father's cousin, to dispel the nagging doubt that in so doing he might have become his own grandpa.

Einstein has a lot to answer for. Thanks to him, scientists, especially those who grapple with great and weighty issues like the origin of the universe, are pictured in the lay mind as elderly men with a shock of white hair like a fright wig. Nothing could be further from the truth.

At its highest levels science, like mathematics and chess, is a young man's game. As they grow older scientists, particularly mathematical physicists, lose the ability to conjure up the dazzling insights of their youth. If a thinker gives the impression of continuing his productive life into late middle or old age it is usually the case that what he is doing is tidying up, elaborating or rationalising the mental leaps made much earlier in his life and, like the best ones in a box of chocolates, saved up for later.

The nimbleness of Einstein's brain was not matched by his hands, which were of the two left thumbs variety, as anyone who ever heard him play the violin will attest. This was a source of deep sorrow to him as he yearned to be able to do experiments well, like his role model Newton, who inexcusably was an experimentalist of genius.

Einstein introduced the expression 'einfühlung', the intellectual's love of the objects of experience, which has promising applications in many areas of bluffing. But not this one.

Michael Faraday (1791-1867)

One of the very greatest of all scientists, perhaps the greatest experimentalist. If you refer to our indicators of success in science, you will find that Faraday wrote the book. He gave his name to various laws, an effect, a unit of measurement, a constant, a piece of equipment, and would have had an element named after him but all the ones known at the time were spoken for.

Faraday came to science quite late in life after a false start as a bookbinder and learned chemistry sitting by Humphry Davy (he of the miner's lamp). Moving to even greater things in physics, he would undoubtedly have won Nobel prizes for both subjects but for bad timing: he died fifty years before the first ones were awarded. As a final gesture, fitting in a man of genuine humility, he turned down a knighthood from Queen Victoria.

Richard Feynman (1918-1988)

Feynman's was a towering intellect, exceeded only by his ego. Among his many hobbies were drawing, bongo drumming, lock picking and theoretical physics. He will ever be remembered for Feynman diagrams, which are best scribbled on the tablecloth in strip clubs. He taught himself the skills of draughtsmanship in order to visit topless bars on the pretext of sketching the girls and the clientele.

This remains the finest excuse yet devised to explain to your spouse why you frequent low dives. Fittingly, he even won his Nobel prize jointly with someone called **Schwinger**.

Galileo Galilei (1564-1642)

One of the seminal figures in modern science, his championing of the views of Copernicus got him into hot water with the Catholic church and he became expert in writing with the fingers of both hands firmly crossed. The story that he tested the law of gravity by dropping his balls off the leaning tower of Pisa is not only inherently unlikely but actually apocryphal.

Sentenced by the inquisition to spend the final years of his life under house arrest, he had the last laugh by smuggling the manuscript of his great book *'Dialogue concerning two new sciences'* to Holland for publication, thus starting the Dutch tradition of bringing out dodgy books that continues to this day.

As father of the scientific method Galileo helped to prise the dead hand of Aristotle from the tiller of science and was one of the principal influences on Newton. These things alone would have guaranteed him a prominent position in the pantheon of scientific demi-gods.

J. Willard Gibbs (1839-1903)

A giant among scientists who had the misfortune to specialize in thermodynamics. No book on science would be complete without a few well-chosen words on that subject so here goes: 'Phooey to thermodynamics.'

Stephen Hawking (1942)

Not only probably the greatest living scientist, but incredibly one who has achieved that status despite having been virtually paralysed for most of his adult life. Understandably, he has had more written about him than Jesus of Nazareth. And now they've made the film of the book of the science of the man.

So the message is 'Bluffers beware'. He is known and blown, and everybody is a Hawking expert. But do not abandon hope. The good professor can still be an inspiration to bluffer and bluffee alike if approached from the right direction.

Everyone has heard of *A Brief History of Time* but what is not so well known is that Hawking has written another bestseller, a collaborative work with **George Ellis** called *The Large Scale Structure of Spacetime,* so unreadable it makes Proust look like Julie Burchill. Its sales have outstripped the number of professional cosmologists, the target readership, to such an extent that it has clearly been purchased by people who not only cannot read it but have no intention of even trying to do so.

Research shows that the average purchaser of *ABHoT* follows a set pattern in his attack on the work. It is done alone and in decent privacy or, failing that, in the presence of one other consenting adult.

The subject checks that there is nothing really good

on television, then settles in a comfortable armchair, the light coming over the right shoulder. Spectacles, if worn, are meticulously cleaned and the blurb and acknowledgements checked one last time to be certain that there is no mistake and there are really no equations in the book (apart from $E=mc^2$ which, like everyone else, he believes he understands).

The first 18 pages or so are read quite briskly but diligently, with some earnest referring back and forth to check knotty points. With an effort, even the discourse on playing ping-pong on moving trains according to Newtonian laws is mastered, and the reader chuckles with the great man at the sad plight of poor old Bishop Berkeley, who thought everything was an illusion.

However, the first grimace, accompanied by a light beading of sweat on the brow, appears with the exposition leading to Einstein's ideas on absolute time on pages 19-20.

The reader rallies gamely when Einstein's famous equation puts in an appearance as advertised, and for a little while he continues to devour the text with a certain rigour. But with the arrival of 'events forming three-dimensional cones in four dimensional space-time', rigour as in painstaking gives way to rigor as in mortis.

The muscles at the angle of the jaw bulge alarmingly and the eyes become dull and staring. The subject starts reading the same sentence over and over again and commences to sigh deeply and repeatedly. He then checks how many pages remain unread and a groan bursts uncontrollably from his lips.

Desperately, he starts to skip-read the remainder of the volume, the eyes flicking left and right, searching desperately but vainly for some crumb of comfort.

The skips get longer and longer, the shuttling of the eyes more and more rapid, until brought to a shuddering halt by a bruising collision with the celebrated final phrase in the text proper '...for then we would know the mind of God'.

At this point the subject stifles a sob, mutters something inaudible about starting to see what Bishop Berkeley was driving at, dons his jacket and stomps out of the house.

Once in the hostelry, however, he starts to perk up and join in the conversation and from then on, on the slightest pretext, he will drag in *ABHoT* and remark casually what a splendid work it is.

If you should happen to be in that bar, seize the opportunity for some reverse, or counter-bluffing. Murmur reverently, "God yes! Hawking's treatment of causality violation in chapter 12 is really magnificent." If the subject agrees with you, especially if he does so more vehemently than really necessary, he is yours, for of course Hawking never mentions causality and the book ends at chapter 11.

Show no mercy. He wouldn't. You should never feel any qualms about hammering a fellow bluffer; your objective is not merely to run with the hare and hunt with the hounds, but to lurk in the hedgerow and machine-gun the lot as they lope past.

Hawking's major contribution to human knowledge lies in the field of black holes. If you are ever cross-questioned about the subject by someone showing alarming signs of interest, our advice is to sidestep.

Tell your interrogator you've never been to Calcutta. If he persists, say, 'Oh, you mean Schwarzschild singularities!' in a tone of amused exasperation. That should throw him off the scent.

If all else fails, sigh deeply and say in a monotone,

preferably without drawing breath at any point, that black holes are hypothetical cosmic features with such intense gravitational fields that nothing can escape from them and they suck in material like giant celestial vacuum cleaners. Nobody is clear how the dustbag gets emptied in this case.

Unlike cornflakes, which come only in large and the giant economy size, black holes can either be small or very small. Mini black holes are one of Hawking's very own concepts and are thought to be left over from the big bang. These gradually lose energy and disappear.

Since nothing can escape from a black hole, including light, they cannot be seen and thus their very existence is conjectural, deduced from theoretical and inferential evidence only. However, since Hawking says they exist it would be a brave person who attempted to deny it.

The irony is that the gallant professor, the supreme cosmologist, is said to have looked through a telescope only once in his life. He got such a headache that he had to have a lie down to get over the experience.

Werner Heisenberg (1901-1976)

One of the onlie begetters of quantum mechanics, he is best known for his deceptively simple uncertainty principle. This states that it is possible to measure the position of a particle or its velocity but not both at the same time. For those with a limited attention span and who find doggerel a useful aide memoire, try, 'Place but not pace, or arse about face.'

Heisenberg is one of the few scientists to have inspired memorable graffiti, *viz*. 'Heisenberg may have been here.'

Sir Fred Hoyle (1915)

A pugnacious Yorkshireman and brilliant mathematician who seems to delight in taking risks with things like his career and reputation. He also cocks a mean snook.

Among the lost causes that he has taken a perverse delight in championing down the years, the best known is the now discredited Steady State theory which he espoused with two famous refugees, **Gold** and **Bondi**.

For a while this theory, which in essence holds that within the expanding universe matter is continuously being created everywhere so as to keep its mean density constant, ran neck and neck with Big Bang. Every time a piece of new evidence contrary to Steady State appears, Hoyle conjures up an ingenious modification to the theory to accommodate it, but in the minds of just about everyone else it has long since ceased to have any credibility.

Hoyle has done a lot of absolutely outstanding work, including the elucidation of the mechanism by which all the elements heavier than helium are formed in the interior of stars, and shares with Lise Meitner the unenviable accolade of the best scientist never to get the Nobel prize.

It seems the earnest Scandinavians fear that if they give recognition to Hoyle's 'serious' work they might be seen indirectly to be condoning his oddball ideas.

We are not certain if there is such a thing as reincarnation, but if there is, Sir Fred will come back as a gadfly.

Lord Kelvin (William Thomson 1824-1907)

An eminent Scottish physicist whose contribution should not be obscured by the fact that he was one of the eminent men of science at the close of the 19th century who said that there was nothing new left to discover. He shares with **Clausius** the dubious distinction of having inflicted upon the world the dreaded second law of thermodynamics.

The unit of temperature, the degree Kelvin, is named after him. It is identical to the degree centigrade but starts at −273.15 instead of square one.

Johannes Kepler (1571-1630)

Scientists are often held to demonstrate elliptical, i.e. deliberately obscure, behaviour, but Kepler's thought processes were literally elliptical. For he it was who demonstrated that the Earth and the other planets go round the sun not in the satisfyingly perfect circles beloved of earlier astronomers, but in ellipses.

Antoine-Laurent Lavoisier (1743-1794)

Prior to the advent of Lavoisier there was a rather charming belief that when substances burned they gave up an invisible, insubstantial material called phlogiston. The French have always relished ruining a good story by the introduction of a few ill-chosen facts and Lavoisier was no exception. He proved that combustion consisted basically of the combination of the material burnt with the all-too-substantial oxygen. He was beheaded during the French revolution, not for this piece of pettifoggery, but for being a tax collector, which seems fair enough.

James Clerk Maxwell (1831-1879)

Bril! ant Scottish physicist who, inspired by Faraday, tidied the ragbag of theories that existed before he came along into a single exquisite theory of electro-magnetism. In turn, he inspired Einstein with his theoretical prediction that light must have a constant velocity.

He also invented Maxwell's Demon, a hypothetical creature capable of opening and closing doors so rapidly that it could segregate molecules.

He used the demon to demonstrate a way of circumventing the second law of thermodynamics but another one of those niggardly Frenchmen, **Brillouin**, eventually found the fatal flaw in this delightful concept.

Sir Peter Medawar (1915-1987)

Medawar was born in Brazil of a Lebanese father and English mother and so naturally went on to become the archetypal urbane, witty, literate Englishman. He was immensely tall, had the flattest feet in Christendom and was one of the finest scientists Britain ever produced, winning the Nobel prize for medicine in 1960 for his work in immunology.

He showed tremendous courage in fighting back from a severe stroke and left a wonderful record of his life and work in the many books he wrote, of which the most noteworthy is his marvellously titled autobi-ography, *Memoir of a Thinking Radish*.

Abbé Gregor Mendel (1822-1884)

Saintly Austrian botanist, Augustinian monk, and pea-sorter, who took time out from his spiritual and administrative duties to found the science of genetics. Regrettably, later analysis by the statistician Fisher has shown that the results of some of Mendel's plant-breeding experiments were just too good to be true, and there seems little doubt they were massaged to fit the theory.

It does not automatically follow that the good friar was the culprit. He relied heavily on assistants in his scientific endeavours and you might charitably assume they told their highly esteemed leader what they thought he would like to hear. Even if Mendel himself was the masseur he suffered enough because fame and recognition came long after his death.

Dmitri Ivanovich Mendeleyev (1834-1907)

An illustrious Russian chemist who invented the periodic table, a systematic classification of the elements. It contained gaps but Mendeleyev was able to predict not only that they would be filled by elements yet to be discovered, but what their properties would be.

His predictive powers certainly came up trumps in the fame sweepstakes by enabling him to have one of his missing elements named after him.

Sir Isaac Newton (1642-1727)

Newton would undoubtedly be in any scientifically-inclined schoolboy's all-time great eclectic cricket team, probably as opening bat. You can open your own innings by opining that Newton's master work, which should be referred to in tones of hushed reverence as 'The Principia' and never by its full title of *Philosophiæ Naturalis Principia Mathematica*, must surely be the most influential single book on science ever written. You are on pretty safe ground here because no less a person than Hawking thinks so too.

Newton's contribution to science was immense, covering such diverse fields as light, colour, the fundamental laws of motion, and gravity. In addition, if ever he found that the mathematics necessary for the full development of his latest idea was not available he would invent that too, e.g. the differential calculus.

He did dabble in chemistry too, but was not too hot at it, which is odd because that is usually the sign of an inferior mind.

Newton made the immense mental leap of realizing that the force which caused the famous apple to fall was the same one that kept the Moon on station circling the Earth. He imposed his will and intellect on nature to produce a beautifully ordered deterministic world-view in which everything wheeled along precisely calculable paths.

This view was to prevail for over 200 years until the arrival of quantum mechanics and relativity showed that the world was not only stranger than Newton knew, but stranger than he could know, given the level of scientific knowledge available to him.

For it is vital to remember that while great geniuses generate inspired flashes of insight that may revolutionise the state of thought at the time, the fact is that even the greatest of them can only produce an idea whose time has come. The greats can get slightly ahead of the game and that is the measure of their genius. But if they are tardy in publishing their findings, other fine but lesser men will discover them independently, such is the pressure generated by an idea that is ready to be born.

Newton was, in the opinion of many, the greatest scientific brain of them all. (Certainly an each-way bet covering Newton and Einstein should ensure the retention of your shirt.) But sad to say, after this it's all downhill, for in truth Newton was a pretty nasty piece of work.

Many exalted men of science before and since have exhibited a certain brusqueness. Still others have been downright rude. But Newton was a pitbull physicist, a Manassa mauler of a mathematician, snapping and snarling at anything that moved. He attacked his mind games and human opponents alike with a mental chain-saw. One doesn't mind one's heroes having feet of clay but Newton was pure kaolin from the neck down.

During his lifetime, he threatened to incinerate his mother and step-father, and conducted vitriolic running feuds with many of the most eminent men of the day. These included **Robert Hooke** of the law of the same name, who once dared to offer some mild criticism of Newton's work and later swapped accusations of plagiarism with him, the Astronomer Royal **Flamsteed**, who perversely declined to see his role in life as supplying data for Newton to build into his latest law, and the German mathematician Leibniz, who not only had the effrontery to arrive at the calculus independently, but had the gall to publish it first. He finally had a run in with a group of English Jesuits who had the misfortune to live in Liège and accordingly were in no mood to take any nonsense from anybody.

This last storm in a teacup so enraged Newton that he actually suffered a complete breakdown. Later he became Master of the Royal Mint, where he turned his ire on the capital's counterfeiters and delighted in having several of them strung up.

He also became president of the Royal Society and used it to his own advantage in the various disputes he continued to engage in. Apart from presiding over meetings of the Royal Society whilst inspecting the inside of his eyelids, he also defended his old university, Cambridge, against attempts by the King to turn it from Protestantism to Catholicism, and represented his alma mater as Member of Parliament. He became the first scientist to be knighted.

It is ironic that among his legacies is Newton's Law of Cooling, for the man himself passed the greater part of his life in a state of self-stoked white-hot incandescence. Quite why has been the subject of intense speculation down the years.

For starters, his father died before he was born, and his mother, perhaps sensing what was coming, promptly remarried and went off with her new husband to raise a brood, leaving Newton with his grandmother for nine long years. And while you are looking for an explanation of his evil temper you could do worse than consider his wig. If you had to go through life wearing a wig as ridiculous and uncomfortable as the one in which Newton is normally portrayed you could be bitter and twisted too.

It is also possible that he was sexually repressed. Quite what his sexual proclivities were is not known, but presumably when it is said that he threw himself into Descartes at Cambridge this is just a figure of speech.

He eventually fell out in a big way with two of his

few remaining friends, the eminent philosopher John Locke and the diarist and stirrer Samuel Pepys. He passed his declining years in tinkering with the great works of his earlier years, expunging from them all reference to anyone who had ever crossed him.

But none of this, harrowing though it may be, seems to explain satisfactorily why he who had so much going for him should have suffered such a curdling of the milk of human kindness. No doubt when he suffered the inevitable decline in his powers it affected him more than the average genius. After all, he had further to fall than anyone else, for he had been the geniuses' genius, the head serag, quite simply, The Man.

We did wonder whether Newton might not have been prey to cognitive dissonance, the inner conflict stemming from incompatible beliefs held simultaneously, since he was a devout, albeit unorthodox, Christian (a closet antitrinitarianist, in fact); but we dismissed this as laughable as, faced with any clash between his own views and those of the Almighty, there seems little doubt which way Newton would ultimately have jumped.

But none of these alternatives offers as convincing an explanation as that the true cause of Newton's choleric mien is that he had the misfortune to be born on Christmas day and so all his life had to endure cheapskate friends and relatives giving him one gift to cover both.

Alexander Pope's lines *'Nature and Nature's Laws lay hid in Night. God said, Let Newton be! and All was Light!'* have been widely misinterpreted. The key is 'Let Newton be!' It should not be read in the sense of, 'Let there be light!' but rather 'Let the blighter alone for Pete's sake' and then All is Clear.

Louis Pasteur (1822-1895)

Chemist and microbiologist who was born in Dole of a
long line of tanners but transcended these potential
handicaps triumphantly to become the most famous
French scientist, giving his name to an effect, an
institute, a genus of bacteria, various bacilli, an
animal disease and a process for preserving food-
stuffs.

Any one of these would have been sufficient to
ensure his place among the immortals of science, but
they all pale into insignificance compared with his
feat in saving the French wine and the British beer
industries by isolating and overcoming the micro-
organisms that plagued them.

Wolfgang Pauli (1900-1958)

Pauli is well known for his exclusion principle, which
sounds as if it should define the criteria for membership of a select club, and that is precisely what it does
do. The potential members in this case are 'matter'
particles whose spin is an odd whole-number multiple
of ½. The club is the state of identicalness.

As normally applied, the principle says that an atomic
orbital containing two electrons of opposite spin is
filled and all the other little electrons must go and
play elsewhere until such time as one of them vacates
it. You may regard the principle as a sort of theoretical
set of carpenter's tools because it facilitated the
construction of the periodic table.

In 1931 Pauli deduced the existence of the neutrino,
which was not actually observed until 1956. Pauli
died shortly afterwards. Presumably he hung on until
he'd had the satisfaction of being proved right.

Max Planck (1858-1947)

The serious bluffer does not deign to make jokes about this German physicist's name, scorning to shoot fish in a barrel. And indeed, Planck's work was no laughing matter, for it was he who, along with Einstein, persuaded science to stop winding up Newton's clockwork universe. This, brilliant though it had been, had ultimately turned out to be a cul-de-sac.

Planck and Einstein kick-started the revolution in physics which spread through the whole of science. Planck took a deep platonic interest in black bodies, and originated his trend-setting quantum theory to explain their heat radiation in terms of its quantization – which simply means that it comes in bite-sized packets called quanta. Eventually it was realized that all electromagnetic radiation is quantised, including light itself.

Planck is also notable for being ahead of his time in other ways. Thus, he became a theoretical physicist before theoretical physics was even invented.

Linus Pauling (1901)

Planck may come after Pauling in the alphabet, but Pauling always has the last word. He is the renegade chemist who took the revolutionary new principles of quantum mechanics and applied them to the study of molecular structure, thereby transforming our understanding of the chemical bond. (Sticking bits of your anatomy together with superglue is only one example of a chemical bond and not a very important one at that, although you might be forgiven for not seeing it that way at the time.)

He also pioneered the idea of resonance hybridization, which might have come to naught had not the Communist Party denounced it as contrary to the principles of dialectical materialism. This immediately made Pauling's name in the decadent West and earned him the Nobel prize for chemistry in 1954.

Pauling applied the stochastic method throughout his career, which means he is a good guesser. Later, when inevitably his intuition dried up, he turned his attention to nuclear disarmament and made such a nuisance of himself that in desperation they gave him the Nobel peace prize in 1962, thus making him the only man ever to receive two unshared Nobels.

Strange to relate, none of this is what Pauling will be remembered for. He will go down as the man responsible for many people's daily consumption of heroic quantities of industrial-strength antioxidant vitamins, ascorbic acid, beta carotene, and vitamin E.

This is based on the belief that they prolong life by mopping up stray free radicals in the system before they can get up to mischief.

They are also now thought to inhibit the oxidation of lipoprotein particles in the bloodstream which helps to stop them gumming up the works.

We prefer to rely on the reducing phenols present in red wine, which are thought to do the same sort of job and not only prolong life but actually make it worth living.

ODD FACTS

These scientific sound-bites are all nuggets of pure gold, hand-hewn from the surrounding hard rock for your predilection and edification. You will be able to make use of them in a variety of ways.

For example, if the opportunity presents itself, you will be able to drop one casually into a conversation, thereby conveying a totally spurious but highly convincing impression that you actually know whereof you speak, which is after all the nub and essence of bluffing.

Then again, if you have the misfortune to be caught by someone who really does know his stuff and you are trapped on the ropes in a neutral corner being pummelled unmercifully, you could always try fcinting with an odd fact chosen quite at random. The more of a non sequitur it constitutes the better, for what you are doing is creating a smoke-screen into which you can disappear until the heat is off.

If ever you are challenged, quite a good ploy is to quote the authority of the European Institute of Economic and Industrial Organizations. Only the occasional crossword freak will realize that the acronym from that nonexistent but plausible body is E.I.E.I.O.

Odd Fact No. 1 – The first triumphant experimental verification of the general theory of relativity was a complete botch. Eddington led a British expedition to West Africa in 1919 in order to make use of a solar eclipse in a part of the sky where there were some bright stars. Observation of these during occultation enabled him to demonstrate that the sun's gravitational field deflected the light from the stars exactly as Einstein had predicted.

However, much later it transpired that the agreement had been a fluke brought about by the cancellation of errors, or the rejection of data that didn't prove the point. Fortunately, by the time this embarrassing realization dawned, later more accurate measurements had amply demonstrated the desired agreement between theory and observation.

No. 2 – Common-or-garden toast is absolutely teeming with free radicals. Fortunately they do not seem to do you any harm, unless you eat more than a hundredweight a day.

No. 3 – The arch-physicist Lord Rutherford said that science can be divided into two categories, physics and stamp collecting. In retribution he was awarded the Nobel prize for Chemistry.

No. 4 – The standard unit of force, the newton, is not defined as the force exerted upon the unprotected napper by a British standard apple falling on it out of a tree from a given height. (If there were any justice it would be, but there is no justice.) On the other hand, coincidentally, it is just about the force required to lift the average apple.

No. 5 – Newton, who was never short of a word or two, especially if they were abusive, only spoke once during all the time he represented Cambridge University in the House of Commons, that natural home of invective. He asked an usher to close a window that was causing a draught.

No. 6 – There are some insects that chirp faster the

higher the ambient temperature, enabling them to be used as a rather inconvenient thermometer. You simply count the number of chirps in ten seconds and that is the centigrade temperature.

No. 7 – Science has a patron saint, Saint Albert, Latin Patriarch of Jerusalem, a man whose presumed piety is exceeded only by his obscurity.

No. 8 – The only unit of measurement more oddly named than the slug-foot-second which is used in British aerodynamics, is the twaddel, a scale named after **W. Twaddel** which is used to measure the relative density of liquids. Frankly, it's wasted on that.

No. 9 – One of Aristotle's oddest dicta was that the human mind reaches its peak at 49. We don't know his age when he came to this conclusion but can hazard a guess.

No. 10 – Kelvin used thermodynamics to show that the sun, using any energy source then known, could not have shone long enough to allow the evolution of intelligence, so Darwin had to be wrong. A little-known geologist, Thomas Chamberlain, turned this on its head and argued that since Darwin was obviously right and thermodynamics couldn't be wrong, it had to be the assumption that the sun relied on known energy sources that was mistaken.

Thus he anticipated the discovery of nuclear fusion without having the faintest idea what it was he was anticipating.

GLOSSARY

You should be careful not to be too precise in your scientific pronouncements but here are a few definitions to be suitably delphic about.

Anthropic principle – The idea that the universe is the way we see it because we are the ones who are doing the seeing. If that sounds silly, try a large Scotch. It still won't sound sensible but you'll feel better about it. There is also a *strong* anthropic principle but you don't want to know about that. It's really silly.

Atoms – What were once thought of as the smallest indivisible component particles of matter until found by **Thomson**, **Rutherford** and **Chadwick** to be made up of subatomic particles – protons, neutrons and electrons. Lots of others have since been discovered which are best thought of as 'and-so-ons' since their names all end in '-on', except for **quarks** *(q.v)*, which being honorary Irish don't have to obey the rules, and **wimps** *(q.v)* which are in no position to object.

Black body – Not something buried in a black hole when it dies but a hypothetical object that completely absorbs incident radiation.

Causality – The principle that causes can be related to effects they produce in the everyday world. In the nightmare world of subatomic particles this principle suffers a breakdown. So will you if you think about it too long.

Chaos – The idea that disordered systems are governed by deceptively simple mathematical relationships. And that chaos can swiftly arise from a neat starting point in just a few steps, with apparently trivial differences in the initial conditions giving totally different results. Accidentally discovered in the '60s by meteorologist **Edward Lorenz** while setting up a computer model of atmospheric behaviour, its roots can be traced back to the early years of this century and the physicist **Henri Poincaré**. How apt that Chaos should be of French provenance.

Chirality – Left or right-handedness in nature. The characteristic smell of oranges and lemons, for example, is due to the right- and left-handed forms of an otherwise identical molecule (limonene).

Conservation of energy – Nothing to do with insulating the loft, but a cornerstone of science until shot down in flames by Einstein who showed that mass and energy are manifestations of the same thing.

Cosmology – The study of the origin and evolution of the universe.

DNA – Deoxyribonucleic acid, the molecule which carries the genetic code which determines heredity. The elucidation of its structure was one of the triumphs of modern science. **Recombinant DNA** is the basis of genetic engineering.

Doppler shift – The apparent change in the frequency of a wave if the source and the observer are in relative motion. Used to be demonstrated by people playing trombones on moving trains before that was made a criminal offence.

Force – Something that operates in fields, but not a

veterinary surgeon. There used to be just two, the gravitational and electromagnetic varieties, believed to operate through the ether. Now joined by two more, the nuclear forces (weak and strong), all are believed to act by the exchange of virtual particles, not faxes. One object of GUTS (q.v.) is to show that they are all manifestations of one fundamental force.

Gibbs free energy – Thermodynamic function any change in which tells of the likelihood of a chemical reaction proceeding. Bluffers know intuitively that it is a snare and a delusion for there is no such thing as free energy any more than free lunch.

Gravity waves – Predicted emissions from massive objects in motion. As yet undetected due to being so weak that even with ultra-sensitive equipment locating them is like trying to find the proverbial thin dime at the bottom of a tall jar of molasses using as a probe a well-cooked stick of rhubarb.

Gypsy – Whimsical name for the J-psi particle, a meson with an unusually long life which accordingly wanders rather a lot.

Hooke's law – That strain is proportional to stress within the elastic limit, i.e. the point where you snap. So obvious to anyone who is married as not to need stating.

Human genome – The human genetic code as recorded by the sequence of base pairs in the DNA; not to be confused with the garden gnome.

Laws of nature – Principles that are believed to hold throughout the universe, the exception being Newton's third law of motion which states that every action gives rise to an equal and opposite reaction. This is true everywhere but the USA, where every action gives rise to a law suit.

Le Chatelier's principle – If you apply a constraint to a system in equilibrium, those changes will take place in the system which tend to remove the constraint you went to all that trouble to apply. It normally refers to chemical systems but we strongly suspect it has universal applicability.

Nano – Prefix for a thousand millionth, from the Greek name for a dwarf; enormous compared with a **femto** which the Greeks didn't have a word for.

Natural Science – the study of objectively measurable phenomena.

Parity – The apparent unwillingness of nature to plump for left or right. **Lee** and **Yang** (two Chinese-American physicists) postulated that some interactions governed by the weak nuclear force might demonstrate a preference and **Wu** duly showed them to be mainly left-handed. Could it be a sinister universe?

Periodic table – The arrangement of the elements into an array in order not of inside-leg measurement but atomic number. With a suitable fudge to allow for the fact that hydrogen is one of a kind, the rest fall into vertical families with similar properties.

Physical Sciences – The hard ones.

Quantum Mechanics – The mathematical formulation of quantum theory which essentially says that energy is discrete, unlike scientists who are indiscreet.

Quarks – Subatomic particles, named by American Nobel laureate **Murray Gell-Mann** after a snatch from Joyce, and distinguished by different 'colours', akin to electric charge but stemming from the strong nuclear force. The successful theory of **Quantum Chromodynamics** is the combination of these colours in different particles.

Second law of thermodynamics – The idea that in a closed system entropy increases, entropy being a measure of disorder. Sounds innocent enough but is about as harmless as an enraged hamadryad. For example, perpetual motion is impossible. Why? Because it would contravene the second law.

Spectrum – The spread of radiation into its constituent frequencies. Visible light is bounded by ultraviolet and infrared, as shown by Newton, whose own behaviour ranged from ultra vires to infra dig.

Super-symmetry – That which when applied to elementary particles gives rise to the splendidly-named correspondent particles selectron, squark, slepton and wino. Try to say that and sound serious.

Taxonomy – Not Italian schedule D, but the science of classification, particularly of biology.

Technology – The prodigal offspring of science that provides the material objects for human comfort and sustenance. Increasingly its cost impedes the progress of science.

Wimps – Weakly interacting massive particles.

Zwitterion – An ion bearing both positive and negative charges; one of nature's oxymorons.

THE AUTHOR

Brian Malpass was born in Stoke-on-Trent, but then so were R.J. (Spitfire) Mitchell, Arnold Bennett and Stanley Matthews and it never did them any harm. He now lives in Marlow.

After spending a large chunk of his childhood in an involuntary study of the tuberculosis bacillus, he wound up at Birmingham University with a First in Chemistry and a Doctorate. When asked what he wrote his thesis on, he replies truthfully, "Fag packets mostly."

The actual subject was the use of radio-tracers in the study of free-radical and anionic polymerisation of substituted acrylates and their interpretation in terms of polar effects, steric hindrance and resonance stabilization. And after all these years he can still say that even with several glasses of his favourite claret taken.

He did in fact make a living as a polymer chemist before drifting into surface science and thence into the management of science and technology. After that, by a series of unplanned sideways and upward lurches, he staggered into finance and ultimately general management of a large public company.

He now seems to have embarked on yet another career as a writer, watched spellbound by his wife, a fellow lapsed chemist, and their two grown-up daughters – who have turned out remarkably well in the circumstances.